Urban Wildlands Fire
Pebble Beach, California

Investigated by: Hugh W. Graham

This is Report 007 of the Major Fires Investigation Project conducted by TriData Corporation under contract EMW-86-C-2277 to the United States Fire Administration, Federal Emergency Management Agency.

 Homeland Security

Department of Homeland Security
United States Fire Administration
National Fire Data Center

U.S. Fire Administration Fire Investigations Program

The U.S. Fire Administration develops reports on selected major fires throughout the country. The fires usually involve multiple deaths or a large loss of property. But the primary criterion for deciding to do a report is whether it will result in significant "lessons learned." In some cases these lessons bring to light new knowledge about fire--the effect of building construction or contents, human behavior in fire, etc. In other cases, the lessons are not new but are serious enough to highlight once again, with yet another fire tragedy report. In some cases, special reports are developed to discuss events, drills, or new technologies which are of interest to the fire service.

The reports are sent to fire magazines and are distributed at National and Regional fire meetings. The International Association of Fire Chiefs assists the USFA in disseminating the findings throughout the fire service. On a continuing basis the reports are available on request from the USFA; announcements of their availability are published widely in fire journals and newsletters.

This body of work provides detailed information on the nature of the fire problem for policymakers who must decide on allocations of resources between fire and other pressing problems, and within the fire service to improve codes and code enforcement, training, public fire education, building technology, and other related areas.

The Fire Administration, which has no regulatory authority, sends an experienced fire investigator into a community after a major incident only after having conferred with the local fire authorities to insure that the assistance and presence of the USFA would be supportive and would in no way interfere with any review of the incident they are themselves conducting. The intent is not to arrive during the event or even immediately after, but rather after the dust settles, so that a complete and objective review of all the important aspects of the incident can be made. Local authorities review the USFA's report while it is in draft. The USFA investigator or team is available to local authorities should they wish to request technical assistance for their own investigation.

For additional copies of this report write to the U.S. Fire Administration, 16825 South Seton Avenue, Emmitsburg, Maryland 21727. The report is available on the Administration's Web site at http://www.usfa.dhs.gov/

U.S. Fire Administration

Mission Statement

As an entity of the Department of Homeland Security, the mission of the USFA is to reduce life and economic losses due to fire and related emergencies, through leadership, advocacy, coordination, and support. We serve the Nation independently, in coordination with other Federal agencies, and in partnership with fire protection and emergency service communities. With a commitment to excellence, we provide public education, training, technology, and data initiatives.

TABLE OF CONTENTS

URBAN WILDLANDS FIRE
Pebble Beach, California
May 31, 1987

Local Contacts:
Robert P. Townsend, Battalion Chief
Pebble Beach Community Services District
Fire Department Forest Lake and Lopez Roads
Pebble Beach, California 93953
408-875-4204

Charles Wilkins, Fire Chief
Pacific Grove Fire Department
600 Pine Avenue
Pacific Grove, California 93950

Brian Weatherford, Fire Control Officer
California Department of Forestry and Fire Protection Central Region
1234 East Shaw Avenue
Fresno, California 93710

Roy Perkins, Ranger in Charge
San Benito-Monterey Ranger Unit
California Department of Forestry and Fire Protection
401 Canal Street
King City, California 93930

SUMMARY OF KEY ISSUES

Issues	Comments
Ignition	Illegal campfire.
Structural Issues	Wood shingle roofs. Brush around homes (negative). Double pane windows (positive).
Terrain/Nature Issues	Dry fuel. Heavy fuel loading. Steep slopes – a classic urban wildlands interface.
Firefighting Problems	Water pressure reduced when area electric power shut off. Lack of access to area of origin. Lack of mobility of units once hoses were laid. Fire storm.
Incident Command System	Used to effect. Problems in communications. Problems with reporting-in procedures.
Evacuation	Difficulty in deciding when to evacuate.

SUMMARY

On May 31, 1987, a fire escaped from an illegal campfire in the Del Monte Forest in Pebble Beach, California. The resulting fire burned 160 acres and destroyed 31 structures causing an estimated damage of approximately $18,000,000. There were 18 injuries, including 15 firefighters and 3 civilians. All were minor, requiring no hospitalization.

The fire spread from the forest into the residential area. Control of the fire in the forest was difficult due to heavy fuel load and low fuel moisture. Structures were located on a ridge above the main body of the fire. A fire storm occurred near the top of this ridge, spreading the fire across the residential area.

The spread of the fire through the residential area was aided by wood shingle roofs, natural vegetation around structures, accumulation of pine needle litter on roofs, and the intensity of the fire.

The area had major woodlands fires in the past and is likely to have more in the future. This is a classic example of the need to be concerned about the urban wildlands interface, and has recently been given higher priority by the U.S. Forest Service and the U.S. Fire Administration.

THE FIRE

The fire started in the heavily wooded Del Monte Forest located between two portions of the S.F.B. Morse Botanical Reserve. (It is therefore called the "Morse" fire.) Officials of the California Department of Forestry and Fire Protection believe that a fire in a trash can located at the campsite radiated heat through the metal to ignite pine needles on the ground of this area.

The campsite seemed to have been used on more than one occasion. No specific evidence as to the ages or number of people using the campsite was available. Evidence found at the scene included mattress box springs, a lounge chair, the metal frame for a camp stool, both empty and full beer bottles, and a metal pole for what may have been a lean-to. In addition, two pits had been dug to contain trash in the area. Both pits contained beer bottles, cans, and other trash.

It appears that the property on which the fire started belonged to the Pebble Beach Company. Campfires or open burning in the area was illegal. The area was enclosed by a cyclone fence, but there were weak places in the fence where access could be gained into the area.

The area of origin was on a hill that faced northwest, with an average slope of approximately 11 percent, but then turning much steeper to an approximately 56 percent slope near the ridge to the southeast where the fire spread. Plants in the area include Monterey Pine, Gowen Cypress, Bishop Pine, Coast Live Oak, Blue Blossom, Manzanita, Huckleberry, Coyote Bush, Pampas Grass, as well as other forbs and grasses. Exact fuel loading is unknown, but estimates vary between 40 and 100 tons per acre. Much of the area of the fire had been kept in a natural state; the use of fire in the area was prohibited, as was any development, with the exception of some unmaintained roads.

Fuel moisture was extremely low due to the fact that rainfall in the area was 50-60 percent below normal for the 1987 season. In addition, rainfall had been below normal for the three years preceding the fire. This reduced fuel moisture would add not only the intensity of the fire, but to the difficulty of control. The heavy growth in the area of origin also made access to the area and fire line construction extremely difficult.

Three major chimneys or canyons led up the slope from the area of origin to Los Altos Drive and Huckleberry Hill. These three canyons, coupled with the steep slope and heavy fuel loading, added to the intensity of the fire and funneled the fire into the residential area across Los Altos Drive.

Smoke was spotted initially by a local resident who called the Pebble Beach Security. They, in turn, contacted the Pebble Beach Fire Station at 1535. Three units were dispatched from the Pebble Beach Fire Station at 1537, including a mini-pumper squad, an engine company, and the duty battalion chief, Robert Townsend. Chief Townsend became the Incident Commander for the fire. This initial dispatch was to check on smoke reported to the fire station. At 1539, the mini-pumper reported smoke visible and estimated that they were approximately one-half mile away from the fire. At 1541, a request for a wildland fire response was made by Chief Townsend. Dispatch and manning levels are preset according to weather readings. The dispatch level for the date of the fire was low and one wildland engine (type three pumper[1]) responded from the Carmel Hill Station. This unit was dispatched at 1542 and arrived at the scene at 1545.

Access into the origin area of the fire was limited; basically, the area could only be reached by foot. The mini-pumper was only able to get within approximately 200 feet of the fire and laid a hoseline into the area of origin. The size of the fire was estimated at approximately one-third an acre upon the arrival of this unit. The mini-pumper expended its 90 gallons of water on the fire, but due to heavy brush and heavy fuel loading was unable to control the fire. The heavy brush and timber limited access and the ability to attack the fire directly. It should also be noted that the only wildland engine immediately available was the one dispatched from the Carmel Hill Station The apparatus dispatched from the Pebble Beach Fire Station was structural firefighting equipment (type one pumpers).

At 1555, a request was made for two hand crews, which normally each contain 15 crew members. At 1602, a request was submitted for two additional engine companies. At 1610, a request was made for an air attack. An air tanker for this area would normally be dispatched from Hollister Airfield; however, this unit was already committed on a fire in the Los Padres National Forest. The only uncommitted air tanker available in the State was dispatched from Chico airfield, approximately 210 miles north of the Monterey Peninsula. At 1615, a request was made for a bulldozer on the fire line. At 1618, a request was made for a helicopter with a water dropping bucket.

---

[1] "Type one" and "type three" are designations for structural and wildland pumpers respectively. The type one has at least 1,000 gpm capacity, and the type three 120 gpm.

At this time, hand crews had still not arrived and the decision was made to start a back fire along a gravel haul road which ran north and south through the fire area. At approximately 1630 to 1635, hand crews arrived on the scene. At approximately 1640, two spot fires occurred about 200 feet up the ridge from the haul road. Equipment on the scene at this time included a mini-pumper, one "type one" engine company, three "type three" engine companies, two hand crews, and one helicopter. Handlines were laid to the two spots above the haul road. Although fuel was continuous to the area west of the haul road and to the area east of the haul road, the terrain changed drastically from the haul road east. The slope to the west of the haul road was approximately 11 percent while the slopes to the east of the haul road in the direction the fire spread was approximately 56 percent.

Mutual Aid Requested – At 1643, an hour after the fire was first reported, the Incident Commander requested a county mutual aid response of two response teams. Each team consisted of three engine companies and a chief officer. Predetermined response teams are set forth in the county's mutual aid plan. One additional fire "overhead" member was also requested at 1643[2]. Pacific Grove Fire Chief Charles Wilkins responded with response team A and proceeded to the incident command post in the gravel pit (see position F on the map in Appendix A). Chief Wilkins was asked by the Incident Commander to act as Sector Commander for the structural portion of the fire, which was designated as Sector C. Chief Wilkins established a sector command post at the intersection of Los Altos Drive and Constanilla Way. He later said there was some mix-up with the response teams: the county communications dispatcher of the day was using a draft matrix (or table) of the response teams on call, and not the matrix which was normally used. This created confusion and may be one reason why equipment from Fort Ord was not requested. (See the Appendix for an example of the response matrix.)

Upon arrival at Sector C, Chief Wilkins found at least three engines from the two response teams already committed with hoselines laid. These engines had taken up position along Los Altos Drive. At least two of the engines were connected to 5 inch lines. The engines set up along Los Altos Drive were in the direct path of the fire and in a very precarious position to defend. Action in this area included attempting to stop spot fires and prevent the spread of the fire up the slopes onto Los Altos Drive. Thus, upon his arrival, Chief Wilkins was not certain what equipment he had available, and some of the equipment was already committed. Not all of the equipment of the response teams had checked into the staging point, in addition to the wrong dispatch matrix being used.

At 1708, the Incident Commander requested two additional water dropping helicopters. During this time, the two response teams were setting up on Los Altos Drive and residents in the area were beginning to water down roofs and areas around their homes. Chief Wilkins later indicated that the use of water hoses and other measures taken by local residents did not seem to impair the water supply at this time.

At 1713, the Incident Commander requested water tenders for the engines located along the gravel haul road. The tenders carried between 2,000 and 3,000 gallons of water. Water supply still was not a problem with the response teams located along Los Altos Drive in Sector C.

At 1721, the Monterey County Dispatchers called the California Department of Forestry and Fire Protection to ask about the possibility of evacuation for the residential area. According to Art McDole, County Emergency Operations Coordinator, they were advised there were no plans for evacuation at

---

[2] "Overhead" includes the incident and sector commanders and staff people.

that time. Mr. McDole indicated that the California Department of Forestry and Fire Protection was called due to the fact that so many calls were being received from citizens in the area.[3]

At 1841, a request for a county mutual aid strike team was made. A strike team consisting of five engine companies with chief officers as leaders was sent to Sector C to work in the residential area. Chief Wilkins indicated that there were still some problems with keeping track of resources, as some members of the strike team got separated en route to the fire and some did not check into the staging point for the sector. Chief Wilkins also indicated that unfamiliarity with the area by the strike team members and the confusing layout of the roads in the residential area probably added to the confusion of the incoming response and strike team members. Street signs in the area are on small wooden posts and are nonreflective.

Evacuation Starts – By this time, three hours after the fire started, some citizens in the area had begun to leave on their own. Security personnel from the Pebble Beach Company were also beginning to secure the area. Spotting (small fires) began above Los Altos Drive and into the residential area. Apparently some county deputies and police officers were also arriving in the area although they had not been officially requested at this point. Pacific Gas and Electric (PG&E) personnel were also in the area.

At approximately 1845, the Incident Commander requested an agency representative from the Sheriff's Office to assist with the evacuation of citizens, and the official evacuation began. This turned out to be about ten minutes before the fire swept over the ridge into the residential area being evacuated. Spotting was continuing to occur above Los Altos Drive into the structures. Although some citizens began evacuation, others continued to attempt to water down roofs and stay with their homes.

Water Press Drops – At approximately 1849, workmen with PG&E cut the power to the residential area. This was standard procedure to minimize danger from energized electric lines to firefighters and residents. The workmen acted after receiving information that a structure was on fire in the area. The cutoff of electric power also cut off power to the pumps for the water tanks of the California American Water Company which were located in the area. The water system consisted of an 800,000 gallon storage tank and a 10,000 gallon pressurized tank. After the power was turned off, the pumps were unable to maintain pressure in the 10,000 gallon tank. With use by the engines, as well as local residents, the tank rapidly lost pressure. This loss of water pressure hampered the fire suppression efforts on Los Altos Drive. Water supply was now a problem in Sector C.

At approximately 1850 to 1855, Sheriff's deputies were attempting to fully evacuate the residential area. Most citizens left, but a few were determined to stay with their properties. The area evacuated was primarily along Los Altos Drive, Sunset Lane, El Bosque Drive, and Costado Road. Approximately 175 to 200 people checked into the Red Cross Emergency Center at the Monterey Peninsula College gymnasium. In addition, at least one area hotel offered rooms free for those forced to evacuate the area. And some residents went to other nearby hotels and motels.

---

[3]There was major community concern after the fire about the evacuation. A report subsequently prepared by Mr. McDole for the Monterey County Supervisors was critical of the evacuation decision and problems with the Incident Command System communications. There is agreement by various parties that there were problems in these areas.

Fire Storm Occurs – At approximately 1855 to 1900 hours, a fire storm occurred in the area along Los Altos Drive at the head of the fire. The fire storm drove the fire over Los Altos Drive and scattered fire throughout the residential area along El Bosque Drive and Sunset Lane. Engines stationed along Los Altos Drive found themselves in an untenable situation surrounded by fire. Chief Wilkins ordered all firefighting operations ceased, equipment removed, and crews to assist with any further evacuation needs of citizens in the area. Equipment was removed and despite the intensity of fire, no major damage was reported to equipment or personnel. Chief Wilkins estimated the fire storm lasted less than five minutes.

After the fire storm stopped, Chief Wilkins ordered the engine companies back into the area. Suppression efforts then focused on directly attacking houses which appeared savable. Numerous roofs were on fire and several houses along Los Altos Drive were already fully involved. Crews also were ordered to direct efforts at exposure control. Water supply at this time was a problem. Water tenders were moved into the area to supply the engines. Chief Wilkins estimated that the majority of the houses lost in the area were ignited during the fire storm or shortly thereafter. He estimated that most of the structural fires were controlled within two to two and a half hours after the fire storm. The residential area was not secured until midnight, May 31, almost eight hours after the fire started. Additional requests for equipment and manpower were made, and the fire was finally "contained" at 1800 hours on June 1, 1987, 27 hours after the fire began. Extensive mop-up was conducted on the fire scene and the fire was reported "controlled" or out at 0800 hours on June 4, 1987. The incident had lasted 3-1/2 days.

CAUSES OF RESIDENTIAL FIRE SPREAD

It appears that the fire spread through the residences because of a number of factors, including wood shingle or shake roofs, pine needles and litter accumulation on top of roofs, dried natural vegetation around the homes, and ultimately the fire storm and increased wind. Although the fire had entered a residential area, it had the aspect of a wildland fire and spread through the wildland-type vegetation between and around the residences. Attempts to maintain a natural type environment in the residential area provided a wildland fuel situation.

The exact cause of the fire storm is not known for certain. It appears that several factors contributed to it, including the heavy fuel loading and the steep terrain coming up to Los Altos Drive. In addition, there were three main canyons or chimneys coming up the slope onto Los Altos Drive. Not only were the fuels heavy in this area, but moisture was low both in the brush and the heavier type fuels. This added to the energy of the fire in and along the area to the west of Los Altos Drive.

Another factor which probably contributed to the creation of the fire storm was the prevailing high winds moving inland from a west/northwesterly direction. This was driving the fire in the east/southeasterly direction onto the slope at Huckleberry Hill along Los Altos Drive. However, firefighters in the area of Los Altos Drive indicated that winds were coming from the northeast. It appears these opposing winds met on the ridge along Huckleberry Hill, and contributed significantly to the creation of the fire storm. The opposing winds may have contributed to an unstable condition above the area, and coupled with the intensity and energy released from the fire, may have been the triggering mechanism for the fire storm. It may also be this opposing northeast wind which prevented the spread of the fire to the east/southeast to more of the residential area. Some spotting did occur east of the main fire and some to the east/southeast. Spotting in these areas came from flying debris of wood shingles or shakes. The largest spot was approximately three acres in size and was located approximately one-half mile southeast of the main fire (D on the diagram attached). Wood shingle material was found in the origin area of this spot. Wood shingle material from the roofs of the houses definitely added to the spread of the fire.

EQUIPMENT AND PERSONNEL USED

The equipment utilized in the fire has been estimated as follows: Two response teams with three engines per team; six "type one" strike teams with five engines per team; four "type three" strike teams with five engines per team; five to six additional engines supplied by local governments: 15 hand crews with 15 people each (nine from the California Department of Corrections and six from the California Youth Authority); nine hired water tenders; two fixed wing air tankers; one air attack coordinator aircraft; eight helicopters (one from the California Department of Forestry and Fire Protection, and seven hired medium and light helicopters, two of which were used for observation and reconnaissance and the rest for water carrying); four dozers (two California Department of Forestry and Fire Protection medium dozers, one hired medium dozer, and one hired heavy dozer); an overhead staff from the California Department of Forestry and Fire Protection estimated at 26; and other agency overhead such as municipal chiefs, estimated at eight people. In addition, a fire information office for contact with the media was staffed with four people during the day and four at night. Finally, two private felling crews with two men per crew were utilized for cutting down trees during the fire.

In all, the California Department of Forestry and Fire Protection estimated 885 firefighting personnel on the fire. Many of these personnel were involved in final control and mop-up of the fire and were not involved in the initial attack or initial spread of the fire.

CASUALTIES AND LOSSES

Fifteen firefighters were treated for minor injuries. Ten were for smoke inhalation, three for eye irritation, one for burning embers in eyes, and one for a head injury. Three non-firefighters were treated for minor fire injuries, two for eye irritation, and one for smoke inhalation.

Some minor damage did occur to fire equipment and apparatus in the area, but exact information on this damage was not readily available.

The total damage from the fire was estimated at approximately $18,000,000. Thirty-one residences were totally destroyed. Eleven vehicles were totally destroyed. Partial damage occurred to six homes, primarily roof damage. Other damage occurred at 14 residences, including fence and vegetation damage. Damage also occurred to three utilities, including PG&E, Pacific Telephone, and Monterey Peninsula Cable TV, whose trailer and microwave power were destroyed.

FIRE DEPARTMENTS IN THE AREA

Staffing of the Pebble Beach Fire Department is done under a schedule A contract between the Pebble Beach Community Services District and the California Department of Forestry and Fire Protection. The Pebble Beach Community Services District owns the facility and apparatus, and the California Department of Forestry and Fire Protection provides the personnel. Apparatus at this station is for structural fire protection, and it has no special wildland equipment or apparatus.

The closest wildland equipment is stationed at the California Department of Forestry and Fire Protection station at Carmel Hill. Normal apparatus at this station includes two type-three wildland engines and one type one engine. This engine is under contract with the California Department of Forestry and Fire Protection, the Pebble Beach Community Services District, and two other Monterey County Districts. All three districts share expenses of this engine and the California Department of Forestry and Fire Protection provides personnel.

The California Department of Forestry and Fire Protection normally increases its staffing during the summer fire season. The Pebble Beach fire occurred prior to the normal fire season and wildland personnel had not yet been increased.

CODES

The Pebble Beach Community Services District has an ordinance requiring class B or better roofing on all new construction and on new roofs in the Pebble Beach area, but this ordinance had been in effect for only approximately 14 months prior to the fire. There was some new construction in the area of the fire, but most of the homes were several years old and were built prior to ordinances requiring the class B roofing. Many of the homes had wood shingles or shakes, and this may even have been required at one time for housing in the area. The Carmel Hill fire station itself has a wooden shake roof that was required at the time it was built.

A Monterey County ordinance requires use of fire retardant roofing materials in the north county area. Other parts of the county's unincorporated areas may not have specific codes requiring roofing materials; however, permits to build are subject to fire district approval in these areas, and fire resistive roofing materials may be required as a condition for obtaining a permit.

State law also requires vegetation to be cleared at least 30 feet from buildings in wildland areas. This is Public Resources Code (PRC) #4291. Information obtained indicates that in the Pebble Beach area, the regulation is enforced only when a complaint is registered. House to house annual inspection for compliance has been discontinued due to lack of manpower. A copy of the Public Resources Code #4291 is included in the Appendix.

No codes requiring residential sprinklers were in effect in the area, and none of the homes destroyed had sprinklers. The majority of them burned from the top down. The effect that residential sprinklers would have in such a situation was undetermined.

DAMAGE ASSESSMENT AND INCIDENT REVIEW TEAMS

The California Department of Forestry and Fire Protection normally reviews all major fire incidents. An incident review team is in the process of reviewing this fire, and a full report will be available later this year.

The California Department of Forestry and Fire Protection also sent a damage assessment team into the fire area to examine damages in detail. This team started work on June 1 and consisted of six people. They looked at physical fire damage, such as charring or burning of structures, other improvements, and wildland vegetation. Their damage survey included for each dwelling its type of exterior siding, roofing, windows, PRC 4291 compliance, and primary vegetation. A damage summary for the incident, the damage assessment worksheets, and the vegetative damage assessment are included in the Appendix.

The summary indicates that of the 31 homes totally destroyed, 22 had shake roofs, four had wood shingle roofs, three had formed steel over shake with no insulation between, and two had composition roofs. For exterior siding, 11 of the destroyed homes had wood siding, 16 had stucco siding, three had redwood and stucco, and one was adobe. For windows, 25 of the homes had single pane windows while five had double pane windows and one had a combination. The PRC 4291 compliance (clearance of brush) was listed as follows: in compliance, four; not in compliance, 27.

The assessment team noted that no structure was lost that had a combination of composition roofing, double pane windows, PRC 4291 compliance, and landscaping.[4] In one composition roof structure which was destroyed, fire ingress was from a deck through a single pane window on the side of the home. At 4051 Los Altos Drive, a single pane window imploded at the rear of the structure with only spot fires in the rear of the yard at the time. The house at 4011 Los Altos Drive withstood fire on its own, with only a small deck fire being controlled by a California Department of Forestry and Fire Protection engine.[5] It should be also noted that it had no eaves and that the exterior glass of its double pane windows had cracked but the interior glass held.

Information obtained from the assessment team seems to indicate that _three major factors contributed to the loss of the homes: roofing material, the type of windows, and PRC compliance._ In particular, the house located 4011 Los Altos Drive was a prime example of fire resistive construction. It was located almost directly in the top of a chimney or canyon on the northwest side of Huckleberry Hill. The structure lay directly in the path of the head of the fire. It had a composition shingle roof, double pane windows, no eaves, and was in compliance with the PRC 4291. It also had a 3-4 foot high masonry wall constructed across its front which may also have helped to divert heat away from it. Some damage did occur to fencing around the structure and some roof damage occurred from burning of pine needle debris on the roof. However, the structure basically withstood the entire force of the fire on its own.

Another structure located at 4059 Los Altos Drive was also near the top of a canyon or chimney. This structure had a shake roof, single pane windows, and was not in compliance with PRC 4291. The structure had stucco exterior and survived with only roof damage. However, an engine stayed with this structure during the fire. In addition, a driveway located in front of the fire may have helped create additional distance between the structure and the main heat from the fire.

Another house located at 4048 Sunset Lane also survived the fire though it had a shake roof. This structure had double pane windows but was not in compliance with PRC 4291. The structure did sustain minor fence damage and smoke damage. However, there was a large space behind the house separating it from the forested area. This probably helped reduce the heat exposure.

The History of the House at 4011 Los Altos Drive – The property at 4011 Los Altos Drive, discussed in part above, was purchased approximately one year prior to the fire by Robert and Marie Whittington. The Whittingtons indicated that they sustained damage to a wooden fence and deck and also some minor roof damage from falling embers. There was some exterior damage, such as cracking of the outside pane of the double pane windows and distortion of the plastic light fixtures. There was also damage to shrubs and landscape around the structure and some minor smoke damage. Other than that, it survived.

---

[4] "Landscaping" here means removal of most native vegetation, and planting of other species around the home. Compared to native vegetation, these plants are generally lower in height, spaced further apart, kept watered, and less susceptible to being ignited.

[5] The address of this house was reported as 4009 in the California Department of Forestry report. 4009 was the legal address, but 4011 was the number on the house.

The Whittingtons indicated that the structure had a composition shingle roof and a stucco exterior. The house was constructed approximately nine years prior to the fire. The original builder wanted a composition roof due to his concerns about mold or fungus buildup on wood shingle roofs which might aggravate allergies of his wife. The undergrowth in and around the house had been removed and cleaned up only a few months prior to the fire primarily to control poison ivy in and around the house. The Whittingtons indicated that prior to the fire, no one had contacted them discussing fire prevention or fuels management around the house. They stated that the fire preventive aspects of the construction of the house were not considered in the original purchase.

The Whittingtons said they left the area prior to any organized evacuation. Mrs. Whittington spotted the fire earlier in the afternoon and called 9-1-1. She was informed that it had already been reported. Later in the afternoon, their son, Mark, called the Carmel Hill Fire Station and was informed to wet down vegetation, walls, and the roof of the house. The Whittingtons, their son Mark, and his family, who live in the house permanently, then went to a local motel to get a room. The first information they received about the destruction of the fire came from the local radio approximately two hours after they left the area.

The Whittingtons feel that the removal of undergrowth around their house played a major role in the safety of the house during the fire. They also felt that the composition roof played a factor, along with the double pane windows. Additionally, Mr. Whittington felt that the wall which was located in front of the house may also have helped contribute to the safety of the home.*

HISTORICAL AND ENVIRONMENTAL CONSIDERATIONS

Historical records indicate that fires have previously occurred in this area of the Del Monte Forest. The last fire in the area was in June of 1959 and burned 62 acres but did not destroy any homes. It began in a remote brushy canyon and spread over the Pacific Grove/Carmel Highway now known as Highway 68 or the Holman Highway. The fire was stopped by approximately 300 firefighters.

Back in 1924 a fire burned 100 acres in this area and a fire in 1904 burned approximately 2,000 acres. No homes were destroyed in either of these fires.

The history suggests that the question is not if a fire will occur but when it will occur in this forested area. Certainly the area is along the Pacific coast and is normally in a high moisture and a high fuel moisture area. However, certain weather factors coupled with the high fuel loading create periods when major fires can occur in the area.

Environmentally, the area is noted for rare plants that grow in unique combinations. The S.F.B. Morse Botanical Reserve was dedicated in 1971 as a means of perpetuating them. The Del Monte Forest is one of only two locations containing native strands of the rare Gowen Cypress. Other rare plants in the area include certain species of Monterey Pines and Bishop Pines.

Fire was a natural element in the original environment of the area. Today, however, the environmental issues have drastically changed. Residents of the area like the natural areas around their homes. The natural setting of the forest combined with the spectacular views of the Pacific Ocean add to the overall aesthetics of the area. Natural vegetation near and around homes is considered part of the aesthetics, as are wood shingle or shake roofs. An attempt is made to maintain as natural an environment as possible around the homes.

---

*There are pictures of this house and others in the vicinity in the file of this investigation.

Prior to the fire, local officials felt that any attempt at fuels management would probably have met with stiff resistance from local residents. Certainly, fuels management such as prescribed burning would have met with resistance not only from the standpoint of the fire itself, but from air quality considerations. Fuels management would have been very advantageous, but given the environmental consideration and the probable response of local residents, it would not have been feasible.

LESSONS LEARNED

1. **Pre-fire planning** for mutual aid and incident management greatly reduced problems which could have occurred in this fire incident. The planning for mutual aid insured fast response times, and the Incident Commander knew the amount and types of equipment which were available. Although there were some relatively minor points of confusion, overall the mutual aid worked well. The mutual aid plan is based on the National Incident Management System.

2. **Communications** is a key element in the coordination of any major incident, even more so where mutual aid and various organizations are involved. It appeared that some breakdown of communication may have occurred between the Sector C Commander and the Incident Commander. There also was some controversy about communications between the Incident Commander and the local county emergency coordinator. Use of designated mutual aid radio frequencies during management of such incidents is a must. Communication must be addressed in the mutual aid plans, and communication frequencies assigned. In addition, the Incident Commander must maintain communication with Division or Sector Commanders and other key emergency personnel.

3. In a mutual aid response, as in any response, Commanders need to take the **time to reconnoiter** the terrain in the fire area, along with observing fuels and exposures. They need to know where to go if they have to pull back. Some of the roads in the residential area in the Morse fire made access in the areas difficult. Major exposures were present. Although it was a residential area, there were many areas of wildland-type fuels continuous through the area. Familiarization with such items is a must to insure proper placement of equipment and personnel. Equipment could have been positioned better in this fire had the Sector Commander been more familiar with the area.

4. **Standard check-in** procedures need to be followed by incoming units. Staging points should be designated and incoming units should check in with them. This insures proper placement of equipment and personnel and allows the Division, Sector, or Incident Commander a full knowledge of equipment and personnel involved in the incident.

5. In a major urban wildlands fire such as the Morse fire, firefighting units need to **maintain mobility** in the fire area. Structural fire units normally locate in one area and fight the fire contained within a structure. The Morse fire is a prime example of an urban wildland interface fire in which, although the wildland fire had burned into a residential area, the fire continued to expand at the normal rate of a wildland fire. The normal spread of a wildland fire dictates that the units in a residential area must be mobile to pick up spot fires and to fight structural fires as they occur with the spread of the wildland fire. It also means they have to be prepared to pick up and move out if necessary.

6. A **critique** is needed after any major incident for all parties involved. The Morse fire is a prime example that mutual aid works; however, as with any major incident, certain points need to be reviewed for future improvement. A critique of the incident by all parties insures improvement of the system and maintains rapport among the mutual aid parties.

7. Firefighters need **psychological support** after an event such as the Morse fire due to feelings of guilt over extensive losses and negative responses by the press and the community. The fire burned through a major residential area and had the potential to destroy an even greater number of homes. Firefighter response in the areas was along standard procedures for firefighting and for conditions under which the incident occurred, but various factors beyond their control led to the severity of the incident. Civilians may not know what is considered standard procedures in a firefighting operation utilized in such an incident. Unfamiliarity with such procedures can create ill feelings and adverse publicity from local media. For example, some citizens who saw an engine company waiting to be assigned became irate that it wasn't in use. Some citizens got angry at crews who took a break. Other citizens tried to direct engines to their homes. Sympathetic support mechanisms must be utilized by the fire service in the face of such responses and feelings. And the community needs to be informed about what measures were and were not taken, and why.

8. **Construction features and construction materials** can reduce the danger from wildland fires in residential areas. Fire resistive roofing can reduce the danger of ignition and also the spread of the fire. The Morse fire showed that double pane windows can prevent the fire from gaining access into the house through window openings. Unprotected or exposed overhangs on structures can allow fire to ignite the structure and gain access into the structure.

9. Home owners or dwellers should **keep their roofs clean** from flammable debris (such as pine needle litter and twigs), in areas prone to wildland fires.

10. **Removal of wild vegetation** around structures can reduce danger from wildland fires. No structures in compliance with PRC 4291 suffered total loss in the Morse fire incident. Again, the house located at 4011 Los Altos Drive is a prime example of compliance with PRC 4291 and reduction of vegetation around the structure. Planted shrubbery was around the house, but the natural undergrowth had been removed and there were no continuous ground fuels up to the structure itself. Structures on either side of this house were totally destroyed and native vegetation ran up against both of them.

11. Tragic events such as the Morse fire incident can be utilized to **direct positive action in fire prevention or awareness.** Review of newspapers in the days following the Morse fire incident indicate that numerous communities were making plans and preparations to avert such disasters. Many cities reviewed and examined fire apparatus and fire protection systems. Information from such incidents can be utilized to develop ordinances or codes relating to construction or construction materials. Newspaper accounts indicated that several local cities used the publicity and emphasis of the Morse fire to spur a ban of fireworks in their cities. Such incidents create an atmosphere in which education and awareness can be capitalized on by local fire prevention and fire education organizations.

CONCLUSIONS

Three major factors which may have reduced the damage were: Safer roofing materials, double pane windows, and compliance with PRC 4291 as far as reduction of vegetation around the structure. No structures were lost that had a combination of composition roofing, double pane windows, PRC 4291 compliance, and landscaping with fire-resistive plants.

The Morse fire points out again that pre-fire planning for mutual aid works. Although some minor points of confusion existed, overall, the mutual aid response was timely and effective. One major point of contention within the overall incident still appears to be evacuation of citizens in the fire area, though no one was injured.

APPENDICES

A. General Area Sketch of the Morse Fire

B. Residential Fire Area

C. Monterey County Area-Mutual Aid Model

D. Mutual Aid Matrix Showing Response for Pebble Beach

E. Public Resources Code 4291

F. Summary of Morse Incident by CDF Damage Assessment Team

G. Damage Assessment Worksheets by CDF Damage Assessment Team

H. Vegetation Damage Assessment by CDF Damage Assessment Team

I. Letter of Support to CDF Employees by Roy Perkins,
    Ranger in Charge of the San Benito-Monterey Ranger Unit

J. County Emergency Coordinator's Report

K. Notes on Slides Taken by Investigator

L. Notes on Photographs Received from the
    California Department of Forestry and Fire Protection

APPENDIX A

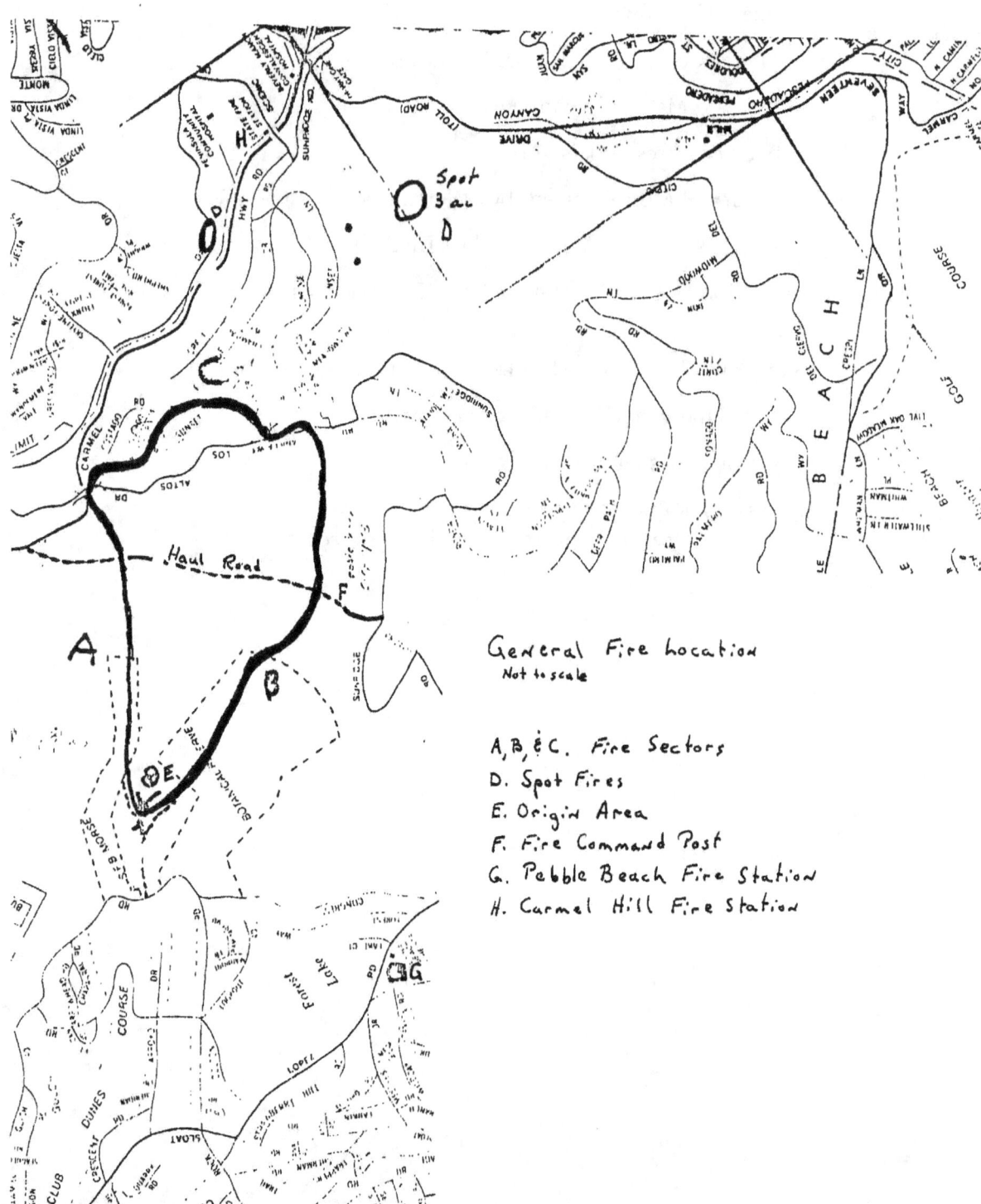

General Fire Location
Not to scale

A, B, & C. Fire Sectors
D. Spot Fires
E. Origin Area
F. Fire Command Post
G. Pebble Beach Fire Station
H. Carmel Hill Fire Station

APPENDIX C

Monterey County Area Mutual Aid Model

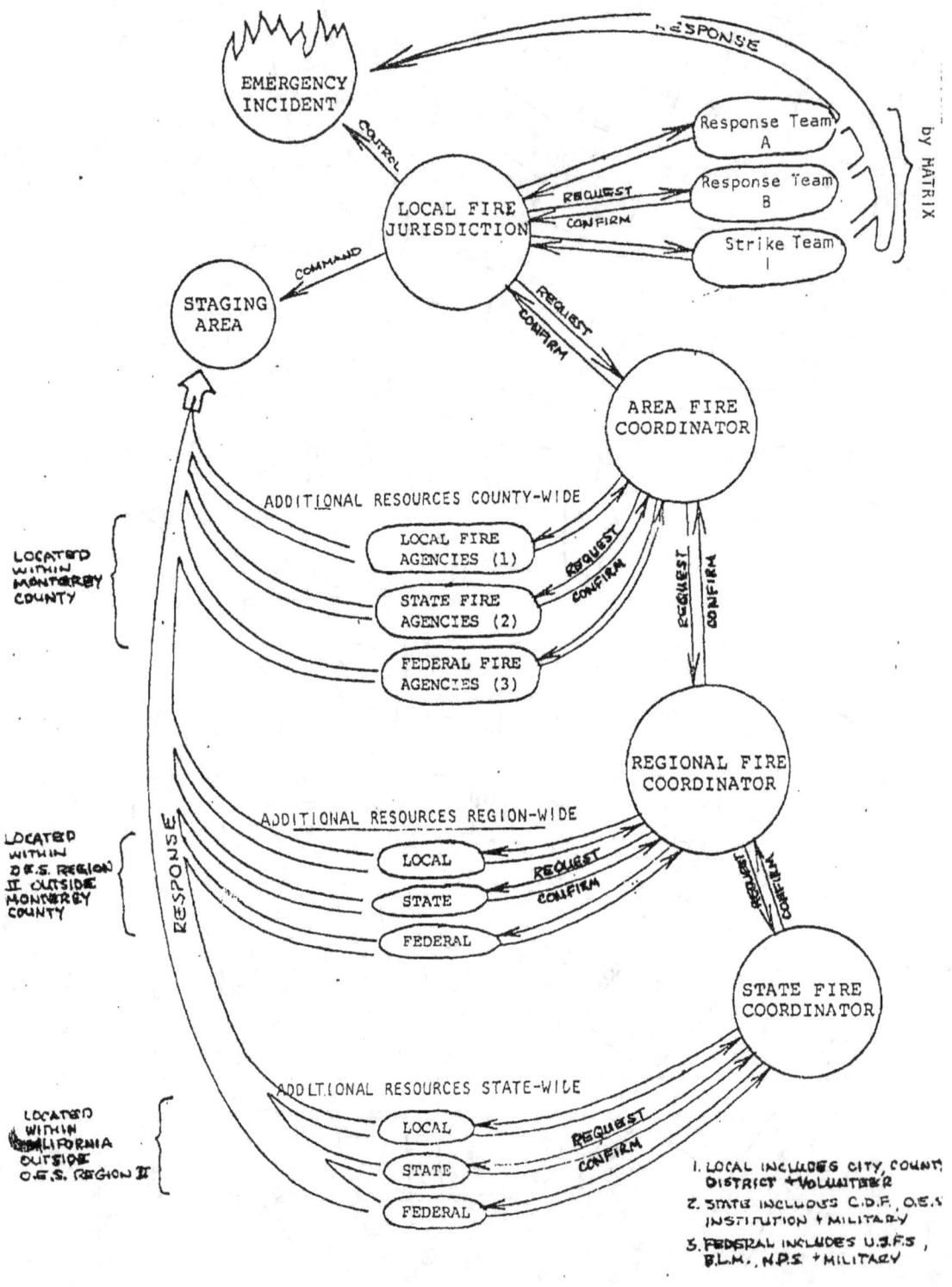

MONTEREY PENINSULA
MUTUAL AID MATRIX

DEPARTMENT REQUESTING MUTUAL AID	RESPONSE TEAM A	RESPONSE TEAM B	STRIKE TEAM I	ADDITIONAL STRIKE TEAMS
CACHAGUA	1 Carmel Valley, Mid-Carmel Valley, CDF-L	1 Carmel City, Carmel Highlands, Salinas Rural	A Monterey, NPGS, Seaside, Fort Ord, Pacific Grove	
CARMEL CITY	2 CDF-L Rio Road, Monterey, Pacific Grove	2 Seaside, Carmel Highlands, Mid-Carmel Valley	B Fort Ord, NPGS, Marina, Salinas Rural, Carmel Valley	RESPONSE
CARMEL HIGHLANDS	3 Carmel City, CDF-L Rio Road, Mid-Carmel Valley	3 Monterey, Mid-Coast, Pacific Grove	C NPGS, Carmel Valley, Seaside, Fort Ord, Marina	ASSIGNED
CARMEL VALLEY	4 Mid-Carmel Valley, CDF-L Rio Road, Carmel Highlands	4 Carmel City, Cachagua, Monterey	D Pacific Grove, NPGS, Seaside, Fort Ord, Marina	by
CDF-RIO ROAD	5 Carmel Highlands, Mid-Carmel Valley, Carmel City	5 Carmel Valley, Monterey, Seaside	E NPGS, Marina, Salinas Rural, Fort Ord, Pacific Grove	AREA
CDF-PEBBLE BEACH	6 Carmel City, Pacific Grove, Monterey	6 Mid-Carmel Valley, NPGS, Seaside	F Fort Ord, Carmel Valley, Salinas Rural, Marina, Carmel Highlands	COORDINATOR
CDF-CARMEL HILL	7 Carmel City, Pacific Grove, Monterey	7 Seaside, Mid-Carmel Valley, NPGS	G Fort Ord, Carmel Valley, Salinas Rural, Marina, Carmel Highlands	
FORT ORD	8 Seaside, Monterey, Marina	8 Pacific Grove, NPGS, Salinas Rural	H CDF-L - CARMEL, Carmel Valley, Salinas City, Carmel City, Mid-Carmel Valley	
MARINA	9 Seaside, Fort Ord, North County	9 Pacific Grove, Monterey, NPGS	I CDF-L - CARMEL, Salinas Rural, Carmel City, Carmel City	

Alert Area Coordinator

APPENDIX E

PUBLIC RESOURCES CODE 4291

REDUCTION OF FIRE HAZARDS AROUND BUILDINGS

Any person that owns, leases, controls, operates, or maintains any building or structure in, upon, or adjoining any mountainous area or forest, brush, or grass-covered lands or land covered with flammable material shall at all times do all of the following:

A. Maintain around and adjacent to such building or structure a firebreak made by removing and clearing away, for a distance of not less than 30 feet on each side thereof or to the property line, whichever is nearer, all flammable vegetation or other combustible growth. This subdivision does not apply to single specimens of trees, ornamental shrubbery, or similar plants which are used as ground cover, provided that they do not form a means of rapidly transmitting fire from the native growth to any building or structure.

B. Maintain around and adjacent to any such building or structure additional fire protection or firebreak made by removing all brush, flammable vegetation, or combustible growth which is located from 30 feet to 100 feet from such building or structure or to the property line, whichever is nearer, as may be required by the State Forester when he finds that because of extra hazardous conditions a firebreak of only 30 feet around such building or structure is not sufficient to provide reasonable fire safety. Grass and other vegetation located more than 30 feet from such building or structure and less than 18 inches in height above the ground may be maintained where necessary to stabilize the soil and prevent erosion.

C. Remove that portion of any tree which extends within 10 feet of the outlet of any chimney or stovepipe.

D. Maintain any tree adjacent to or overhanging any building free of dead or dying wood

E. Maintain the roof of any structure free of leaves, needles, or other dead vegetative growth.

(f) Provide and maintain at all times a screen over the outlet of every chimney or stovepipe that is attached to any fireplace, stove, or other device that burns any solid or liquid fuel. The screen shall be constructed of nonflammable material with openings of not more than one-half inch in sire.
(g) The State Forester may adopt regulations exempting structures with exteriors constructed entirely of nonflammable materials, or conditioned upon the contents and composition of same, he may vary the requirements respecting the removing or clearing away of flammable vegetation or other combustible growth with respect to the area surrounding said structures.

No such exemption or variance shall apply unless and until the occupant thereof, or if there be no occupant, then the owner thereof, files with the State Forester, in such form as the State Forester shall prescribe, a written consent to the Inspection of the Interior and contents of such structure to ascertain whether the provisions hereof and the regulations adopted hereunder are complied with at all times

CHIMNEY
SPARK ARRESTER

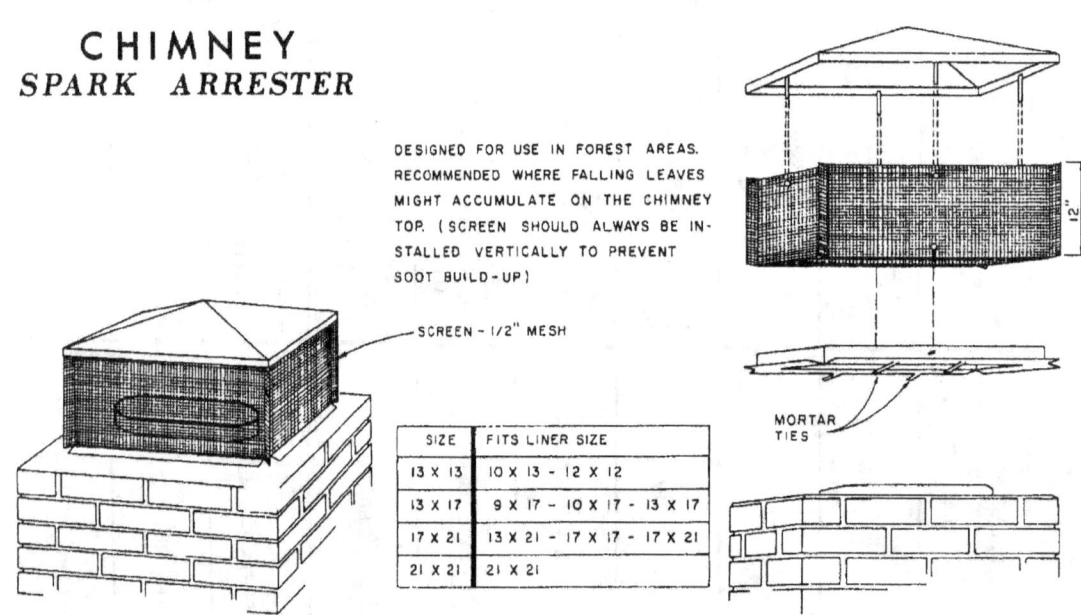

DESIGNED FOR USE IN FOREST AREAS. RECOMMENDED WHERE FALLING LEAVES MIGHT ACCUMULATE ON THE CHIMNEY TOP. (SCREEN SHOULD ALWAYS BE INSTALLED VERTICALLY TO PREVENT SOOT BUILD-UP)

SCREEN - 1/2" MESH

12"

MORTAR TIES

SIZE	FITS LINER SIZE
13 X 13	10 X 13 - 12 X 12
13 X 17	9 X 17 - 10 X 17 - 13 X 17
17 X 21	13 X 21 - 17 X 17 - 17 X 21
21 X 21	21 X 21

Fire
HAZARD REDUCTION
For your Safety and Protection

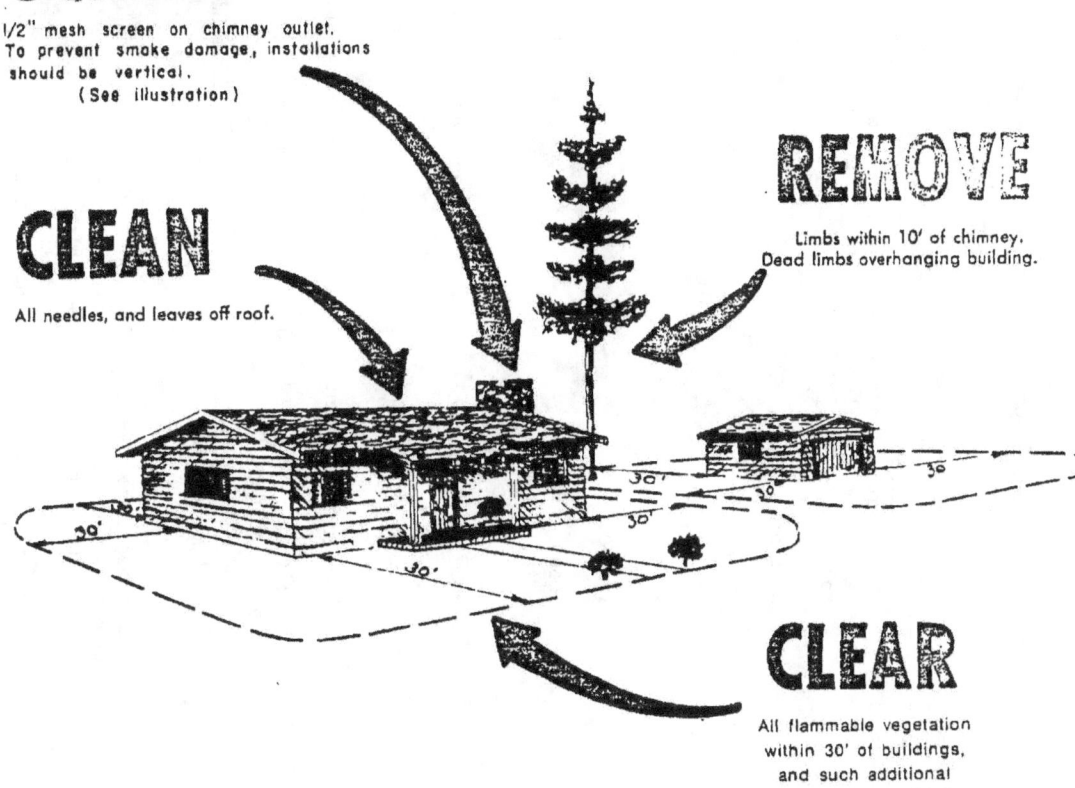

SCREEN
1/2" mesh screen on chimney outlet.
To prevent smoke damage, installations
should be vertical.
(See illustration)

CLEAN
All needles, and leaves off roof.

REMOVE
Limbs within 10' of chimney.
Dead limbs overhanging building.

CLEAR
All flammable vegetation
within 30' of buildings,
and such additional
clearance, up to 100 feet,
as may be directed.
(See reverse side Para. B)

See reverse side for illustration of screen installations.

STATE OF CALIFORNIA
DEPARTMENT OF FORESTRY

19

SUMMARY MORSE INCIDENT

	Total Loss Homes: 31	Partial Damaged Homes: 6	Other Damage: 14 (Fence & Vegetation Primarily)	Utilities Damaged: 3
Roofing:				
Shake	22	5	2	PG&F
Wood Shingle	4		3	Pacific Telephone
Formed Steel over Shake	3			* MPTV
(no insulation between)				
Flat Composition		1		
Tile			1	
Composition	2		8	Vehicles: 11

(Flat composition roof fire needles only, all others roof damage.)

Exterior Siding:			
Wood	11	4	6
Stucco	16	2	7
Redwood & Stucco	3		
Adobe	1		1

Windows:			
Double Pane	5	2	6
Single Pane	25	4	8
Both	1		

PRC 4291 Compliance:			
Yes	4	2	8
No	27	4	6

Primary Vegetation:			
Natural	8	0	0
Landscaped	1	1	4
Combination	22	5	14

* Note : MPTV was not in compliance with PRC 4291.

Note: No structures were lost that had a combination of composition roofing, double pane windows, PRC 4291 compliance and landscaping. One composition roof structure destroyed - fire ingress was from deck through window on sic of home (single pane). At 4051 Los Altos a single pane window imploded at rear of structure with only small spot fires in rear yard at that time. 4009 Los Altos withstood fire on its own with only small deck fires being controlled by a CDF engine. It also should be noted that it had no eaves and the exterior glass in the double pane windows cracked, but the interior glass held.

Appendix G Continued

DAMAGE ASSESSMENT WORKSHEET

Address	Exterior Siding	Roofing	Windows Single or Double	PRC 4291 Compliance	Primary Veg. N-Natural L-Landscaped C-Combination	Structural $ Damage	Contents $ Damage	Misc. $ Damage	Vehicle $ Damage	Extent of Damage-Veg Total & Partial Structural
4024 Sunset	Wood	Shingle	Single	N	C	159,000	111,300	3,000		Total Loss
4028 Sunset	Stucco	Shake	Single	N	C	300,000	300,000			Total Loss
4029 Sunset	Stucco	Shake	Single	Y	L	99,873	74,905			Total Loss
4032 Sunset	Wood	Shake	Single	N	N	261,200	182,840	106,120		Total Loss-Priceless art works.
4035 Sunset	Wood	Shake	Single	Y	C	175,175	131,381			Total Loss
4040 Sunset								200,000		MPTV-Metal Vaults and Microwave
4044 Sunset	Stucco	Shake	Double	N	C	175,000	125,000			Total loss-Estimated val
4048 Sunset	Stucco	Shake	Double	N	C			15,000		Minor fence damage & smo
4055 Sunset	Wood	Shake	Single	N	C	218,000	131,250	17,500		Total Loss
4056 Sunset	Wood	Shake	Double	N	C	190,000	1,000			Total Loss
4059 Sunset	Stucco	Shingle	Single	N	C			15,000		Fence 6 smoke
4064 Sunset	Wood	Shingle	Single	N	C	171,000	149,000			Total Loss
4068 Sunset	Stucco	Shingle	Single	N	C			15,000		Fence 6 smoke
4072 Sunset	Stucco	Shake	Double	N	N			15,000		Veg. & fence & smoke
4080 Sunset	Stucco	Shingle	Single	N	C			15,000		Gazebbo
4004 El Bosque	Adobe	Shake	Single	N	C	175,000	125,000			Total Loss-Estimated val
4008 El Bosque	Redwood	Steel o/Shake	Both	N		239,000	6,500		3,700	Total Loss
4011 El Bosque	Stucco	Shake	Single	N	C	135,000	67,700			Total Loss
4012 El Bosque	Stucco	Shake	Single	N	C	200,000	200,000			Total Loss
4016 El Bosque	Stucco	Steel o/Shake	Single		C	116,300	58,150	34,890	2,000	Total Loss
4021 El Bosque	Adobe	Flat Comp.	Single	Y	C			15,000		Minor fence damage & smo
4025 El Bosque	Wood	Comp.	Single	Y	C	175,000	125,000	15,000		Minor fence damage & smo
4031 El Bosque	Wood	Shake	Single	N	N	145,000	66,000			Total Loss-Estimated val
4032 El Bosque	Stucco	Shingle	Single	N	C	103,000	5,000			Total Loss
4036 El Bosque	Wood	Shake	Single	Y	C	111,000	55,500	11,100		Total Loss
4044 El Bosque	Stucco	Shingle	Single	N	C	25,000				Roof fire
4045 El Bosque	Wood	Shake	Double	N	C			15,000		Fence & smoke damage
4048 El Bosque	Wood	Flat Comp.	Single		C			15,000		Minor damage-burned and broken fence and smoke
4052 El Bosque	Wood	Comp.	Single	Y	C					Roof fire-interior damage
4054 El Bosque	Wood	Shake	Single	Y	L	50,000		15,000		Minor damage-fence & smo
4058 El Basque	Wood	Flat	Single	Y	C			15,000		
Pacific Bell										Phone Service Repair
PG&E								50,000		Electric & Gas Service R

Appendix G Continued

DAMAGE ASSESSMENT WORKSHEET

Address	Exterior Siding	Roofing	Windows Single or Double	PRC 4291 Compliance	Primary Veg. N-Natural L-Landscaped C-Combination	Structural $ Damage	Contents $ Damage	Misc. $ Damage	Vehicle $ Damage	Extent of Damage-Veg Total & Partial Structural
4009 Los Altos	Stucco	Comp.	Double	Y	C			50,000		Veg-cleanup-broken windows paint
4013 Los Altos	Stucco	Comp.	Single	N	C	184,607	138,455		23,000	Total Loss
4017 Los Altos	Stucco	Shake	Single	N	C	55,000	27,000		11,895	Total Loss-Under insured
4029 Los Altos	Stucco	Comp.	Single	N	N	175,000	175,000		10,000	Total Loss
4037 Los Altos	Wood	Steel o/Shake	Single	N	N	182,120	136,590			Total Loss
4041 Los Altos	Stucco	Shake	Single	N	N	275,000	50,000			Total Loss
4049 Los Altos	Redwood/Stucco	Shake	Single	N	C	153,800	153,800	15,380		Total Loss
4051 Los Altos	Redwood/Stucco	Shake	Single	N	N	275,000	100,000			Total Loss
4055 Los Altos	Stucco	Shake	Single	N	N	239,500	119,750	71,850		Total Loss
4059 Los Altos	Stucco	Shake	Single	N	C	25,000				Engine 4673 stayed with structure during fire. Smoke and small roof fire.
4067 Los Altos	Stucco	Shake	Single	N	N	146,000	80,300	14,600		Total Loss
4071 Los Altos	Wood	Shake	Single	N	C	350.000	140,000			Total Loss
4077 Los Altos	Stucco	Tile	Double	Y	L			15,000		Minor fence damage & smoke
4026 Costado Pl	Redwood	Shake	Double	N	C	137,500	137,500	13,750	2,000	Total Loss
4030 Costado Pl	Redwood/Stucco	Shake	Double	N	C	175,000	125,000			Total Loss-Estimated value
4032 Costado Pl	Wood	Flat Comp.	Double	Y	C	25,000				Minor damage Roof fire needles
4038 Coatado Pl	Stucco	Shake	Double	N	C	184,589	101,524		3,000	Total Los;
4046 Costado Pl	Wood	Flat Comp.	Double	Y	L		15,000			Minor damage-Fence & smoke
4050 Costado Pl	Wood	Comp.	Double	Y	L			15,000		Minor damage-Fence & smoke
4052 Costado Rd	Stucco	Shake	Single	N	C	25,000				Roof fire
4085 Sunridge	Wood	Shake	Single	N	C	25,000			25,000	Roof fire / Total loss

5 Vehicles not otherwise noted

APPENDIX H

Vegetation Damage Assessment

I. Introduction

On May 31, 1987 an illegal campfire escaped its confines and ignited a major brush and timber fire in the hills of the community of Pebble Beach, burning most of the vegetation totally. Plant species in this area consist in part of Monterey pines, Gowen cypress, Bishop pine, Coast liveoak, Blue blossom (ceanothus spp.), Manzanita (Arctostaphylos spp.), huckleberry, Coyote bush, pampas grass and numerous forbs and grasses. Much of the area has been kept in a "natural" state, excluding both fire and development with the exception of some roads. The fire consumed approximately 160 acres of vegetation as well as 31 homes before being controlled.

II. Area Description

The fire area lies on a northwest facing aspect with an average slope of 13 percent, with the steepest slopes occurring at the upper end of the fire, near Huckleberry Hill. Three general vegetation areas have been identified for the purposes of this assessment. The first includes the area of origin upslope, roughly to where the main haul road traverses the slope. It is an uplifted ocean terrace and has an impervious hard pan, or "caliche" layer which has led to the retarded development of plants growing on it, commonly referred to as a pygmy forest. The second area extends upslope from main haul road to Los Altos Drive just below the ridge of Huckleberry Hill. The soil profile is better developed resulting in more normal vegetation growth. The third area occurs up from Los Altos Drive including the housing and developed area of Huckleberry Hill. Natural vegetation occurs in this area along with a mix of ornamental species where landscaping has occurred.

III. Damage Assessment

A. Pygmy Forest

Probably 95 percent of the living vegetation in the pygmy forest area, within the fire perimeter was totally consumed by the fire. The overstory consists primarily of Monterey pine. The understory contains a mixture of the other species listed above. Although most of the foliar material is gone, it is likely that by the end of the first growing season (+ August 1988) the area will show ample signs of recovery, sprouting species should be re-established uniformly over the burned area. Monterey pine, which releases its seed in response to fire will have an excellent seed bed and should also show good signs of re-establishment.

B. Normal Forest Area

The forested area on the better developed soils was damaged similarly, although not quite so completely. Again, most of the understory was consumed. Roughly 70 percent of the Monterey pine crowns were desiccated by heat convection and/or actually burned to the point that the trees will not survive. Resprouting and seed germination should show good progress toward re-establishment by the end of the first growing season in this area as well.

C. Huckleberry Hill

The area extending from Los Altos Drive through the developed housing contains a mixture of natural vegetation and introduced ornamentals used for landscaping. Mature Monterey pine forms most of the overstory canopy in this area. Most of these trees were killed by the fire. The understory vegetation, on the vacant lots and in many cases both front and backyard landscaping consisted of live oak, huckleberry and Manzanita. It, too, was almost completely consumed by the flames. Between sprouting and seeding these species should show good signs of re-establishment by the end of the first growing season. The pines will also re-establish themselves, probably to the point of overstocking, depending on the amount of viable seed available for release at the time of the fire. Many of the ornamental species located in and adjacent to yards for landscaping are lost and have no vegetative and or seeding method of reproduction, unlike the natural species of the area which have evolved with fire. They will have to be replaced at the option of the individual landowners, many of whom also lost their homes.

Special consideration should be given to the fire damaged and killed Monterey pine in the developed areas. Monterey pine is a shallow rooting species. With most of the crowns burned and the understory vegetation gone, many of these trees may be susceptible to wind throw by high winds when the ground is saturated by winter rains.

IV. Summary

In summary, probably 75 percent of the overstory vegetation consisting primarily of Monterey pine was killed by the fire. In all probability, at least 50 percent of the remaining live trees will succumb to either insects or disease within the next two years. Close to 90 percent of the existing understory consisting primarily of live oak, Manzanita and huckleberry were burned completely. Revegetation will probably occur naturally, with obvious signs of it by the end of the first growing season. In all probability the area will not see a shift in species composition; however, it will take many years for the area to achieve the look it had prior to the fire. The last major fire in this area occurred somewhere around 1958 or 1959, approximately 30 years ago. Vegetation in this area is indicative of the way the present burned area will recover.

Prepared by Daniel H. Ward, Forester I

APPENDIX I

Memorandum

To : To All Employees
     San Renito-Monterey, Ranger Unit

Date : June 9, 1987

Telephone: ATSS ( )
                  (408) 385-5412
                  Green 387

          San Benito-Monterey Ranger Unit
From: Department of Forestry

Subject: 8000 FIRE CONTROL
          Morse Fire) BEU-775
          May 31, 1987

We read about these things in Newspapers, watch them on TV

We read our training manuals and professional magazines.
We plan, train, drill and meet to coordinate our efforts.

We do not prepare ourselves for IF it will happen;
We prepare ourselves for WHEN it will happen because we all know that
eventually it will .

It happened to us May 31 , 1987 at 15:35. A fire that started out as a
routine "duffer" in a short period of time changed into a violent. major
disaster.

Many you responded to the fire. You all performed well against
insurmountable odds. Those of you that had to remain in your stations or
move up for Cover were giving your support; we knew that you wanted to come
to our aid. Those of you who supported us from King City Headquarters were
of equal importance.

We lost 31 homes and 11 vehicles. We will never have a count on the
numbers we saved, but there were many.

Most importantly we lost no lives and suffered only a few injuries.

We can all stand tall knowing that we did all that was humanly possible.

Do not let those that criticize us force us into a different strategy. We
must continue our preplanning. Make an on-scene plan and execute it.
Waver only when the fire or conditions dictate.

Remember we are still the best!

ROY A. PERKINS
Ranger in Charge

MONTEREY COUNTY

EMERGENCY OPERATIONS

(408) 422-3660 - P O BOX 1883 - SALINAS CALIFORNIA 93901

ART McDOLE
COORDINATOR

June 22, 1987

To: Board of supervisors
From: Art McDole
Subject: Report on Morse Fire

Tuesday, June 2, your Board voted to; (a.) instruct the Chair to write a letter to Chief Roy Perkins, California Department of For estry and Fire Protection, requesting a meeting with me to discuss the Morse Fire, and (b.) to instruct me to investigate the actions taken at the fire and report to your Board.

It is my belief that the Board does not expect me to render a report or an opinion on the cause of, or the efforts to control the fire, I do not feel qualified to make judgement on these items which will be left to experts in that field. In conjunction with the Sheriff and his officers, I have made a detailed investigation on those aspects of the fire which I feel relate to the actions cf our county departments. This report will reflect the findings.

BACKGROUND AND SCOPE OF INVESTIGATION

Investigation was made of the radio and telephone logs which are maintained by the Monterey County Communications Department and those made by the California Highway Patrol and the Pebble Beach Security Company. The information from the CDF logs, released by Chief Perkins at the meeting of Friday, June 5th was also used. I was in attendance, and participated in that meeting. Further information was gained from interviews with victims, deputies and others who were at the fire scene, and from Pebble Beach Security Company personnel.

A briefing was held at the Monterey Peninsula College on Monday, June 8th. This was attended by several county personnel, including myself. This meeting centered on the actions taken at the reception center.

I was at the Command Post at the Pebble Beach Security Office during the fire from about 1900 hours until 0230 hours the next morning. My observation of the actions taken there are reflected throughout this report.

In response to the letter from the Chair, a meeting was held June 15th from 0900 hours until 1200 hours in my office. In attendance were:
Sheriff Bud Cook

26

Appendix J Continued

Sgt. Joe Grebmeier, Sheriff's Department
Sgt. Mike Kanalakis, Sheriff's Department
Lt. John Crisan, Sheriff's Department
Chief Roy Perkins, CDF
Assistant Communications Director Jerry Verwolf
Art McDole, Emergency Operations Coordinator

The purpose and intent of the meeting was stated and it was explained that an attempt is being made to determine what, if any action or lack of action could be identified, and if any changes could be made which would improve our response in any similar future situation.

SIGNIFICANT TIME ELEMENTS

The first report of smoke was from a citizen in Ocean Pines to Pebble Beach Security at 1533 hours, Pebble Beach Security sent a unit to investigate and called CDF at 1535 hours. The first CDF unit was on scene at 1539 hours. Citizens began calling 911 ten minutes after the first call.

At 1608 the Monterey Communications Center notified the Sheriff's Department and at 1610, informally notified the California Highway Patrol (CHP).

The first request from CDF for mutual aid came at 1642, an hour and nine minutes after the first report. However, during this time CDF had dispatched several of their own units, air support, hand crews and a bulldozer.

At 1844, three hours and eleven minutes after the first report, CDF requested the Sheriff to report to the scene for evacuation.

Seven minutes later, at 1851, the fire was reported to have crossed the top of Sunset with many houses on fire.

In determining the significance of these time elements, it is necessary to describe the communications systems.

COMMUNICATIONS CENTERS

The California Department of Forestry and Fire Protection operates a 24 hour communications center in King City. This is their central dispatch and control for both Monterey County and San Benito County.

Monterey County operates two consolidated communications centers, one in Salinas and the other in Monterey, The one in Monterey serves the Pebble Beach area and is-the dispatch and control center for the Sheriff's units in that area. 911 calls received in the Monterey Center if intended for CDF, are directly transferred over a single dedicated line. During the course of the fire approximately 500 such calls

- page 2 -

Appendix J Continued

were received. As many of these calls as possible were transferred,to CDF via this dedicated line. The balance were answered and completed in the Monterey Center. Off duty dispatchers voluntarily responded to the Center, and staff was doubled to handle the extremely high number of telephone calls and radio traffic during the peak hours cf the fire.

All CDF field units communicate with the King City center on their own channels. All Sheriff's units communicate with the Monterey Center on their own channels, The California Highway Patrol communicates with their center on their own channels and the Pebble Beach Security Company has their own dedicated channel. None of these agencies are able to directly communicate with each other, except certain CDF units can go to the Pebble Beach Security Company system if they so desire. The Sheriff's units have scanning monitors which enable them to hear other units, and many of the CHP cars do as well. All chief officers of CDF also have scanners.

Communication, between the Monterey 911 Center and the King City CDF Center, if of an emergency nature, is generally handled over the same single dedicated line that is used to pass the 911 calls.

As will be shown, there was no coordinated field command center, and virtually all communications from field units passed through the various separate agency center and was relayed to the other centers as deemed advisable.

ON SCENE OR FIELD COMMAND

Virtually all fire operations, and CDF operations in particular, handle large incidents using the "Incident Command System" The basic design of this system is to establish a field command with an Incident Commander (IC). The responsibility of this individual is to direct, track and coordinate all response units. If confined to a single service, this is accomplished through the use of various command and general staff officers. Coordination with other agencies is through representatives from those agencies, operating with a Liaison Officer on the Command staff of the IC. The location of the Incident Command Center is determined by the nature of the incident and may be at the actual scene, or at a nearby site if deemed advisable.

All fire incidents are under the control of the fire department having jurisdiction. In this case it was CDF, as the area is classified by state law as a "State Responsibility Area". The fire-fighting activity and request for support services was at their discretion.

Appendix J Continued

OTHER AGENCY INVOLVEMENT

While the Sheriff's Department is responsible for all aspects of law enforcement in the Pebble Beach area, Pebble Beach Company employs its own security force. Their function is to administer gate entrances, patrol the Company's property, and generally render information and assistance to residents, property owners and visitors. Preliminary investigations of minor reports concerning Pebble Beach property and other types of details where the proximity of the Pebble Beach Security force could result in a quicker response are sometimes made by this force.

The Sheriff's Department and the Pebble Beach Security historically work closely together on all types of incidents. While a Sheriff's unit was dispatched in the early stages of this incident, to make an area check, it is probable more units would have been dispatched to a similar incident where there was no resource equivalent to the Pebble Beach Security force. This was brought out in the investigation, but did not materially affect either the action or the outcome.

Pebble Beach Security units were present at all stages of the fire, from the first report until the present time. They were effective in keeping their office informed of the situation, provided traffic control, and assisted in the evacuation. I have not identified any conflict between their forces and that of any other agency, and, to my belief, they played a vital and effective role.

The California Highway Patrol was requested to assist in traffic control, and were very effective in their stations at the gates and along Highway 68. Large crowds of onlookers were attracted by the dense clouds of smoke, and there was heavy traffic. The Monterey Police Department assisted at Highway 68 and Skyline and the Pacific Grove Police Department handled Highway 68 and Presidio.

EVACUATION

In the event of evacuation, the principal responsibility lies with the Sheriff. The <u>decision</u> to evacuate lies with those having expertise with the type of incident. For example, in a flood, this would be the Flood Control people. In a fire, it is the Fire agency. However, in any event, it should be a joint decision involving all concerned.

In the Morse Fire, there was no planned evacuation. As shown by the log, CDF requested the Sheriff units to report to the Granite Pit for evacuation at 1844 hours and seven minutes later at 1851 the houses on top of Huckleberry Hill were on fire and it was an emergency situation, with people fleeing for their lives.

Appendix J Continued

The logs do not indicate any prior request for, or consideration of evacuation, prior to 1844 hours, even though structures on Los Altos had been reported on fire as early as 1639 hours.

On arrival at the scene, the Sheriff's units attempted to evacuate persons both in and very near the fire and those for about three blocks away. They used the loudspeakers on the patrol cars and also went door to door. It was necessary to forcibly evacuate some persons.

Investigation shows that certain Pebble Beach Security Officers assisted in the evacuation, and that three off duty officers, Pat Homan, Joe Panetta and John Panetta, were evacuating persons from burning homes and were probably responsible for saving several lives. It is reported that other citizens rendered valuable assistance in the evacuation process.

COMMAND POSTS

At approximately 1900 hours an effort was made by the Sheriff's Department to establish a coordinated command at the Pebble Beach Security Office. Representatives of the Sheriff's Department, the Pebble Beach Security and Art McDole, Emergency Operations Coordinator were present. There were also SPCA personnel present to handle any animal or pet problems. At the same time, fire operations were being controlled by CDF from a pickup truck parked nearby in the gravel pit area.

This command at the Pebble Beach Security Company offices was attempting to coordinate the final stages of the evacuation, arrange for assistance at the reception center at the Monterey Peninsula College and to assess the situation. This was made very difficult due to lack of direct communication with the CDF IC. The best information from the field was coming from the Sheriff's deputies and from the Pebble Beach units via their individual radio systems. Telephone contact as well as radio contact was established with the Monterey Center, but they were unable to obtain virtually any useful information regarding the status of the fire from CDF at any time during the incident. Chief Robert Townsend, the IC, came to the Pebble Beach Center for a short time, but no permanent liaison was established.

Reports were being received of fire across Highway 68, and there was concern over the potential need to evacuate the Community Hospital and the Skyline Convalescent Hospital. No prognosis was available from the Fire IC.

MAJOR IDENTIFIED PROBLEMS

It does not appear that the Incident Command System

Appendix J Continued

was ever fully established, nor did it function in the prescribed manner. There was no single coordinated command post established at any time.

. There was a severe lack of communication between the Fire IC and the other agencies. This was due in part to limited radio capabilities, and to the fact that no liaison or coordinated command was established.

. No joint agency discussions were held during the entire course of the incident, even though all agencies involved communicated on a one - on - one basis from time to time,

. The Monterey Center did not receive information from CDF on the progress or potential course of the fire, which could be given to the field units.

. The decision to evacuate was not made in a timely fashion.

. Evacuation was accomplished as a frantic reaction, rather than a planned operation, and residents were never informed in advance that evacuation might be necessary.

. The decision to establish a reception center at Monterey Peninsula College was not made by the Sheriff's Department or the Emergency Operations Coordinator, nor were these agencies directly informed that was being done until the decision had been made.

. Many residents were instructed by citizens to go to the Del Monte Shopping Center and there was no organized plan for handling them at that location.

. Complaints were received from residents that they were unable to ascertain if their homes had been lost for many hours after the fire had been contained.

SUMMARY

In spite of these deficiencies and problems, there were many positive aspects to the fire.

. No lives were lost, nor were there any serious injuries.

. The Monterey Peninsula College worked out well as a reception center due to the response and activities of both the Carmel Chapter and the Monterey County Chapter of the American Red Cross. College personnel were very effective in their response and in the manner in which they assisted. The Explorer Scouts rendered invaluable aid in directing persons to the center'.

. Numerous food establishments and hotels were both prompt and generous in their response to the evacuees. They were

Appendix J Continued

fed and offered accommodations at no charge in a very short time.

. The Salvation Army responded with their mobile canteens without being called, and provided excellent food service to the workers.

. Pacific Bell and P. G. and E. responded in a very positive and effective manner during, and immediately following the fire. Gas and electricity were shut off for safety reasons and restored within a remarkably short time frame.

. A State of Local Emergency was proclaimed during the fire by the Emergency Operations Coordinator and response from the State Office of Emergency Services was both prompt and positive.

DECISIONS AND ACTIONS TAKEN AT THE JUNE 15TH MEETING

These decisions were made and the resultant actions were established at the June 15th meeting by mutual agreement of Roy Perkins, Chief, CDF, Sheriff Bud Cook, and Art McDole, Emergency Operations Coordinator.

1. The Sheriff has established a policy to respond deputies to every fire in the unincorporated area to assess the situation and assist as appropriate.

2. CDF will keep County Communications currently advised of the situation on all wild fires where structures may be endangered.

3. In any fire where there is perceived danger to a number of residences or persons, a coordinated command will be established between Fire and Sheriff at an early stage, if possible. If it appears there is a potential requirement for an evacuation, the Emergency Operations Coordinator will be notified. A joint decision will be made regarding notification to residents of the possible need to evacuate. Every attempt will be made to commence actual evacuation in timely fashion.

4. County Communications will acquire portable, multi-channel radio equipment, suitable for use at a coordinated command post, and will deliver it to the scene as required or requested.

5. If evacuations become necessary, they will be performed by the Sheriff and coordinated by the Emergency Operations Coordinator. Efforts will be expended to establish a reception point, track all persons passing through, and keep them informed to the extent possible, of the status of their homes. The American Red Cross will be notified as soon as it is decided to evacuate

Appendix J Continued

and establish a reception center.

CREDIT

I would like to give particular credit to Sgt. Joe Grebmeier,
Sheriff's Department and to Jerry Verwolf, Assistant
Director, Communications Department, for the many hours spent
researching tapes and assisting in other phases of this
investigation.

Respectfully submitted.

Art McDole, Emergency Operations Coordinator

APPENDIX K

Notes on Slides Taken by Investigator

The slides are described by their roll number and number on the roll – R1x9 is slide #9 on roll #1.

Slides with an asterisk have also been made into photos.

*R1x9 -   A view looking northwest at the origin area of the fire. The yellow ribbon visible in the center of the slide surrounds the illegal campsite. The fire originated from a barrel being used to contain fire inside this illegal camp area.

*R1x8 -   A view looking northwest from the area of origin of the fire. Notice the heavy fuel accumulations in the area. Notice also that the vegetation still has most of its leaves in this area where the fire backed away from the origin area of the fire. This heavy debris and fuel in the area limited access to the fire.

R1x7 -   A view looking east/southeast in the direction the fire traveled from the origin area. Note again the heavy fuels in the area and the heavy burning which occurred near the origin area. Some crowning occurred just to the southeast of the origin area.

R1x11 -   A view of one of the many signs in the area of the fire. Notice this is a sign designating private property, no camping, hunting, lighting of fires, or trespassing. These signs are maintained by the Pebble Beach Company owners of the property.

R1x6 -   A view looking at the fuel west of the origin area of the fire. Note that the roads which appear clear now were cleaned during the fire operations. Again, notice the heavy fuel loading in the area.

*R1x10 -   A view looking southeast along the southern perimeter of the fire. Again, the road was cleared during the fire operations. Notice the heavy fuel accumulations to the right of the slide indicating the fuel loading at the time of the fire.

R1x4 -   A view looking at fuels along the western perimeter of the fire. This is a view looking west/northwest toward the origin area of the fire.

R2x20 -   A view looking at the fuels along the southern and southwestern perimeters of the fire. Notice heavy fuel loading and in addition, heavy accumulations of pine needles in the understory. Numerous dead trees were located also in this area, apparently from insect kills.

R1x3 -   A view looking south along the haul road which the suppression crews attempted to back fire. The fire spotted over this haul road initially in two spots approximately 200 feet above the haul road.

*R1x2 -   A view looking east/southeast from the haul road up the slope of Huckleberry Hill. Notice the heavy growth of Monterey Pine and heavy fuel loading underneath. This is one of the canyons or chimneys which came out on Los Altos Drive. This particular chimney was located directly in front of 4011 Los Altos Drive. The address provided in much of the official references to this structure was 4009 Los Altos Drive, but the property address was posted as 4011 Los Altos Drive.

*R1x15 - A view looking west/southwest from Los Altos Drive down the canyon or chimney in front of 4011 Los Altos Drive. Notice the intensity of the fire indicated by the complete burning of all materials in the area. Drowning also occurred along this portion of the ridge.

*R1x5 - A view of the gravel or granite pit which was located on the western portion of the fire and which served as incident command post for the fire. The western perimeter of the fire was located along the east side of this gravel or granite pit.

R1x12 - A view of the intersection of Los Altos Drive and Constanilla Way. This intersection served as the staging area for Section C.

R2x10 - A view looking south/southeasterly at the house located at 4009 Los Altos Drive. A fire engine stayed at this location during the fire and protected this house.

R2x8 - A view looking west/southwest from the second turnout spot on 17 Mile Drive. This was located just opposite from 4011 Los Altos Drive. Notice heavy fuel loading in and around this area.

R1x14 - A view looking south/southeast along Los Altos Drive. This was the area where response teams A and B were located when the fire storm came up the side of Huckleberry Hill.

R2x9 - A view looking east at vegetation in the area of 4013 Los Altos Drive. Notice heavy accumulation of fuels in and around the house. This house was totally destroyed.

*R2x11 - A view looking north along the north side of the structure. Again, notice the heavy accumulation of natural fuels up to the area along the west side of the structure. Again, this would have been in the area where the fire storm first hit Los Altos Drive.

R1x13 - A view looking northeast at 4011 Los Altos Drive. No heavy fire damage was in or around the area. A chimney or canyon from below was located directly in front of this structure. Again, this is the area where the fire storm first hit Los Altos Drive.

R1x17 - A view looking southeast at 4011 Los Altos Drive. Again, notice the heavy damage in and around the structure but again, the structure remained intact. Note that the fence visible in the left center of the photograph was destroyed in the fire. In addition, a wooden deck on the structure was destroyed. Notice there is no overhang on the structure and the structure had a composition shingle roof. The exterior is stucco.

*R2x12 - A closer view of the structure. Note that the street address of 4011 is visible on the exterior of the property. Note there is some vegetation up to and around the structure. However, there is no continuous fuel in and around the structure.

R2x14 - A closer view of the wall which was located in front of the structure. This wall may also have helped to deflect the fire and would have also prevented the spread of the fire directly in front of the structure.

R2x6 - A close view of the landscaping in and around the house itself. Notice there are some small shrubs in the area, but there is no undergrowth around the shrubs and there is no continuous fuels up to or around the structure itself.

R2x7 -   A view looking at the shrubbery in front of the structure. Again, there is distance between most of the shrubbery and there are no fuels in between the shrubbery. Cracking did occur to the double pane windows along the west side of the structure which were exposed to the brunt of the fire and fire storm.

R2x17 - A view looking west at the back portion of the structure. Some minor roof damage was reported but as can be seen; there is no major damage to the structure.

R1x16 - A view looking north/northeast along Los Altos Drive. This is the area just to the north of 4011 Los Altos Drive. At least two vehicles were totally destroyed in this area.

R1x24 - A view of one of the structures that was destroyed in the fire. Notice that burn patterns indicate that the fire started in the upper portions of the structure and came down. As is common throughout the area, fires mostly started in the roof areas and the structure burned from the top down.

R1x18 - A view looking north at the structures destroyed along the ridge top of Huckleberry Hill. This would have been in the area lying between Los Altos Drive and El Bosque Drive.

R1x19 - Another view of a structure destroyed in this area.

R2x15 - A view of the structure located at 4001 Costado Road. This would be across the corner from the intersection of El Bosque Drive and Los Altos Drive. Fire damage did occur to the southeast of this structure, but notice the green lawn in and around the structure and the large space in front created by the intersections of the three roads, Los Altos Drive, El Bosque Drive, and Costado Road.

R1x20 - A view looking at the house located at 4048 Sunset Lane. This structure did have a shake roof, but also had double pane windows. There was only minor fence and smoke damage done to this structure.

R2x18 - A closer view of this structure. There was a large space to the back side of the structure. This coupled with the fact that an engine was placed here during the time of the fire probably accounts for the lack of damage to the structure.

*R2x5 - A view of some of the typical spot damage to wood shingle or wood shake roofs around the area. Note the covering over an area where a hole has burned through the roof at this house.

*R2x3 - A view looking at an area where the structure to the left has a composition roof while the structure to the right had a wood shingle roof and is totally destroyed.

R2x16 - A view looking at the house with a tile roof and wood siding. The structure had no major damage.

R2x4 -   A view looking at the structure located at 4021 El Bosque Drive and had an exterior of adobe. The roofing was flat composition with single pane windows. The structure was in compliance with PRC 4291. The structure had minor fence damage and smoke damage.

*R2x2 - An example of some of the aftermath of the fire. This house had minor roof damage and previously had a wood shingle roof. As can be seen, the house is now having the roof replaced with tile.

R1x25 - An example of some of the accumulation of pine needle litter on top of structures in the area. This is probably the extreme case in the area, but indicates that heavy accumulation of pine needles can occur in such areas. These pine needles are on top of a wood shingle roof.

R1x22 - A view of the 800,000 gallon water storage tank located on top of Huckleberry Hill. Note there was heavy wildland fuel accumulations around this storage area. Again, the area was left in a natural state as possible to provide pleasing aesthetics to the area.

R1x21 - A view of the 10,000 gallon pressurized tank in the water system.

R2x22 - A view of the common road signs in the residential area where the fire occurred. Notice these are of a nonreflective type and again are somewhat small and tend to blend into the surrounding area. Such signs, although aesthetically pleasing in such an area, can hinder rapid response of units unfamiliar with the area.

*R1x2

*R1x15

*R2x12

*R2x5

*R2x3

*R2x2

*APPENDIX L

Notes on Photographs Received from the California Department of Forestry and Fire Protection

Photo A is a view looking east at the overall area of the Morse fire. The red arrow in the lower center of the photograph indicates the spread from the origin area which was below the red arrow. The black arrow in the upper right of the photograph indicates the area in the gravel pit where the Incident Command Center was located. The black arrow in the left center of the photograph indicates the haul road which extended north and south across the center portion of the fire. The haul road is visible extending from the black arrow in the left of center in the photograph through the upper center of the photograph to the gravel pit in the upper right. The black arrow in the upper left of the photograph indicates the turnout which was located directly in front of 4011 Los Altos Drive. This is a major area in the upper left of the photograph where the fire extended up to the top of Huckleberry Hill and into the residential area.

Photo B is a view looking east at the top of Huckleberry Hill. The black arrow to the left of center in the photograph indicates 4011 Los Altos Drive. This gives some idea as to the fuel loading around this structure, as well as the damage that occurred around the structure. This also gives some idea as to the fuel loading throughout the residential area on top of Huckleberry Hill.

Photo C is a view looking westward from the top of Huckleberry Hill in the direction where the fire came into the area. Again, in this photograph you can see the steep terrain coming up to the west side of Huckleberry Hill. Also, the residence located at 4011 Los Altos is visible in the lower right corner of the photograph. Note that the structures that were totally destroyed are visible to the left of this structure.

Photo D is another view looking westward from the top of Huckleberry Hill. Again, the residence located at 4011 Los Altos is visible in the left center of the photograph. This shows a general view of the fuel loading in the area, as well as the damage which occurred in the area.

Photo E is a view looking eastward at the house located at 4011 Los Altos. The house is located just to the left of center in the photograph but is nearly invisible due to the heavy trees in and around the area.

Photo F is a view looking westward on the top of Huckleberry Hill. The water storage tank is visible in the lower left corner of the photograph. The street visible running through the center of the photograph from left to right is Sunset Lane. Again, this photograph gives some idea of the fuel loading through the residential area and the natural vegetation which was scattered through the area, as well as the damage which occurred in the residential area.

Photo G is a view looking northward along the top of Huckleberry Hill. The red arrow in the center of the photograph indicates the house which was located at 4059 Los Altos Drive. Note the greener color of crowns of trees to the left of this structure. It appears that the fire was not as intense in this area as in other areas onto the north along Los Altos Drive. As has previously been mentioned, an engine company stayed with this house during the fire. The red arrow in the upper right of the photograph indicates the approximate location of the house located at 4048 Sunset Lane. Sunset Lane is the road which is visible in the extreme right of the photograph leading from top to bottom in the photograph. Los Altos Drive is the road which is visible in the left center of the photograph going from lower left to upper center.

Photo H is a closer view of the structure located at 4059 Los Altos Drive. Again, notice the greener crowns of trees to the left and upper left of the structure. Again, it appears that the fire was somewhat less intense in this area.

Photo I is a view looking southward at the structure located at 4048 Sunset Lane which survived the fire. Notice the structure is in the upper center of the photograph and a large open area extended to the back side of the structure. It appears there was a major area of separation around the house from the timber in the area. Again, a good view of the fuel load and the natural vegetation located in the area is visible in the lower portion of the photograph.

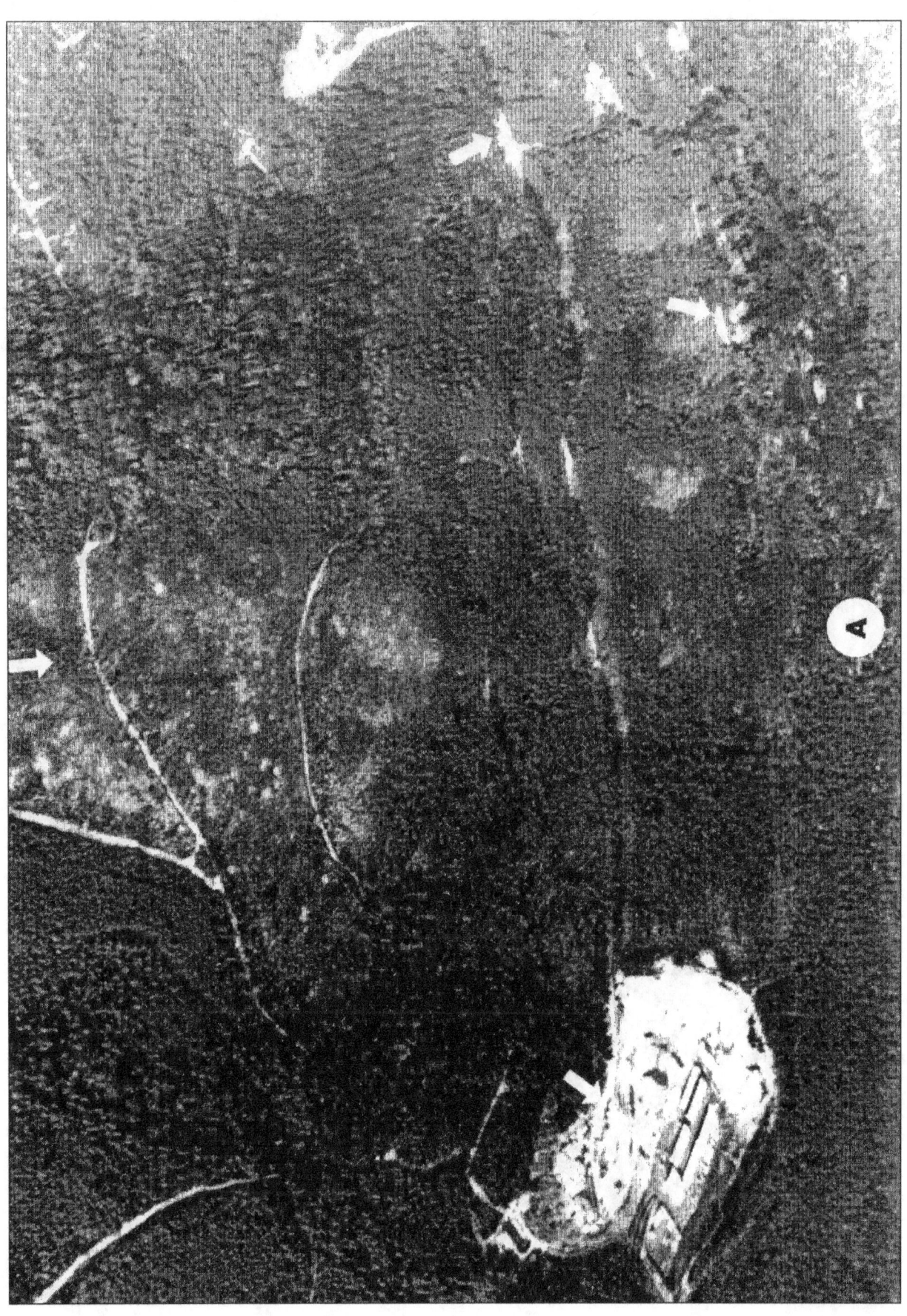